Experiments on Air Pollution

Without air we cannot survive. Clean air is essential for the good health of animals, plants and people. But, every year, 1.5 million tonnes of smoke are discharged into the atmosphere over the UK alone. Smoke from factory chimneys and from homes pollutes the atmosphere and contaminates the air we breathe. Motor cars are another source of air pollution. In Tokyo, Japan, the air pollution from motor car exhausts is so bad that policemen on point duty have to wear gas masks!

Air pollution is difficult to identify. The particles of dust in the atmosphere are often too small to be seen but the effects are obvious – dirty buildings and dead vegetation. This book shows you how to identify, examine and combat air pollution. Its twenty experiments are all graded on a five-point scale according to degree of difficulty. One star represents the easiest experiments, and five stars the most difficult. There is a comprehensive glossary and a further reading section.

ABOUT THE AUTHORS

IVOR WILLIAMS has a BSc in Chemistry, a Diploma in Education, and an MEd. He has been a teacher in secondary and higher education for sixteen years. Currently, he is senior lecturer in chemistry, environmental science, and education at City of Liverpool College of Higher Education.

DOUGLAS ANGLESEA has a BSc in Agricultural Botany, a Diploma in Rural Education, and an MSc. He has been a teacher in secondary and higher education for nineteen years. He is currently teaching in higher education with particular reference to environmental pollution and microbiology.

On next page Fumes pour out of a Bessemer furnace at a steel plant in Workington, England.

Experiments

on

Air Pollution

D I Williams and D Anglesea

Experiments in Pollution and Conservation

Experiments on Air Pollution

Experiments on Land Pollution

Experiments on Water Pollution

Projects in Conservation

ISBN 0 85340 564 6

Copyright © 1978
D I Williams and D Anglesea.
First published in 1978 by
Wayland Publishers Limited
49 Lansdowne Place, Hove,
East Sussex BN3 1HF, England.
2nd impression 1981.

Phototypeset by Direct Image, Hove.
Printed and bound in England by
The Garden City Press Limited, Letchworth

Contents

I Pollutants in air

Introduction

Air is vital to our survival. Its contamination is a serious matter. Although the early humans were able to tame fire they could not tame smoke and other side-effects of burning. As industry grew the problem of smoke pollution from the use of coal became worse.

Today most people live in crowded urban areas. Fuel is used to heat homes, and to move goods about the world. Mills, smelters, electricity generating plants, chemical plants and refineries pour out endless unpleasant substances into the atmosphere.

Some air pollutants are removed by nature. For example, particles of soot settle by gravity. The pollutant gas, sulphur dioxide, is washed out by the rain. Other gases are taken in by plants and trees. But in all cases there are bad side-effects. Buildings, furnishings and clothes get dirty from soot deposits. Sulphur dioxide dissolved in rain eats away the fabric of buildings and makes the ground too acid for plants to grow; plants are killed when they absorb pollutants.

Some of the major by-products of combustion (e.g. carbon dioxide) are not a direct health hazard. But they may have serious long-term effects on climate. They may heat up the earth's surface and eventually melt the ice caps.

Dust from smoke in the atmosphere may reduce summer temperatures, by scattering the sun's energy away from the earth. The earth's animals and plants are then robbed of light and sunshine. Other by-products of industry are poisonous substances like fluorides, lead compounds and carbon monoxide. All of these are poisons which damage health.

In many big cities in winter the contaminants are trapped at ground level by layers of warmer air above. This is called inversion, because normally warm air tends to rise. An inversion causes contaminants to build up at ground level, creating smogs. Smog contains many irritants, including high concentrations of sulphur dioxide. This gas irritates the nose and throat and may cause bronchitis. It also kills plants, and has killed all the lichens which used to live in inner city and industrial areas.

Clearly air pollution has a vast effect on our environment. Pollution has grown alongside wealth. The problem is no longer local: everyone is affected. Governments will soon need to put money and manpower in to fighting this difficult problem.

On facing page Young people protest against pollution by cars at the opening of the New York Automobile Show in 1971.

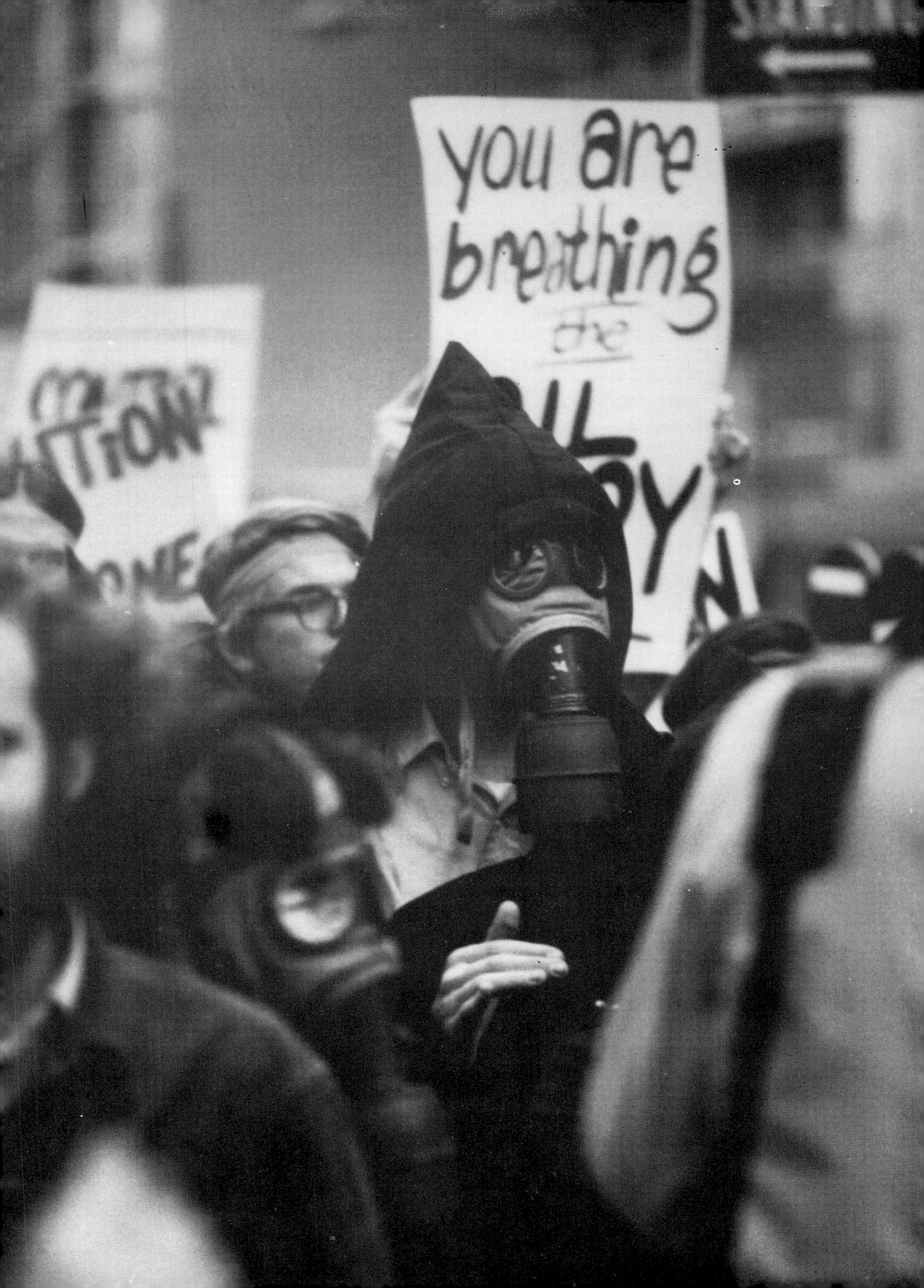
you are
breathing
the

(1) Measuring the oxygen content of air *

Table 1. Constituents of the air

(a) Invariable component gases of dry carbon dioxide-free air

Gas	Formula	% by volume
Nitrogen	N_2	78.110
Oxygen	O_2	20.953
Argon	Ar	0.934
Neon	Ne	0.001818
Helium	He	0.000524
Methane	CH_4	0.0002
Krypton	Kr	0.000114
Hydrogen	H	0.00005
Nitrogen oxide	N_2O	0.00005
Xenon	Xe	0.0000087
Total		99.9997647

(b) Variable components

Gas	Formula	% by volume
Water vapour	H_2O	0 to 7.0
Carbon dioxide	CO_2	0.01 to 0.10, mean 0.034
Ozone	O_3	maximum 0.000007

(c) Contaminants

Gas	Formula	% by volume
Sulphur dioxide	SO_2	up to 0.0001
Nitrogen dioxide	NO_2	up to 0.000002
Ammonia	NH_3	trace
Carbon monoxide	CO	trace, but up to 0.002 at street level in busy towns

Let us begin our study of air pollution by looking at the composition of ordinary pure air. This will give us a basis for comparing the results of the experiments about polluted air which come later.

Pure air is colourless and without smell. It contains enough oxygen to support the life of plants and animals. People at the seaside often say they can smell ozone because the air is so pure. This cannot be true; ozone is very poisonous. If you could smell it, death would soon follow! Seaside air is usually free from soot contamination. The fresh tang comes from the salt water, drying seaweed, tarred ropes, mud and fish.

If the atmosphere had more oxygen in it than at present, fires would be very hard to control and almost impossible to put out. Fortunately, atmospheric oxygen is mixed with nitrogen. There is roughly four times as much nitrogen as oxygen in the air; this is the ideal mixture for supporting biological life. The amounts of oxygen and nitrogen do not vary, so they give us a stable atmosphere. But this could change as pollution increases.

Pure air also contains a small amount (about 0.03%) of carbon dioxide gas (used by green plants to make food), up to 1% of rare gases, mainly argon, and a good deal of water vapour. Have you ever noticed the water vapour trails made by high-flying aircraft?

Diagram 1 Apparatus for measuring the oxygen content of air.

What you will need
A pneumatic trough or washing-up bowl
A tall narrow glass or measuring cylinder
Some steel wool (the sort used for
cleaning pots and pans)
A small flat wooden stick with
cotton thread attached
An ink marker or strips of sticky paper
Several small flat stones

What to do
Place the stones in the bowl and add
water to a depth of 10 cm. Using the
thread, tie the moistened steel wool to
the top of the stick. Place the stick and
steel wool upright in the jar, with the steel
wool at the bottom. Now turn the jar
upside down, and balance it on the stones
in the pneumatic trough. Mark the position
of the water level on the side of the jar.
Record the water level at daily intervals.
As the steel wool rusts, oxygen is
absorbed from the air and water enters
the jar to take its place. When all the
oxygen has been used up the water level
ceases to rise. We can then work out the
rough proportion of oxygen in the original
air mixture.

That is, % oxygen =

$$\frac{\text{rise in water level}}{\text{initial height of air column}} \times 100$$

$$= \frac{y}{x} \times 100 \text{ (see diagram 1)}$$

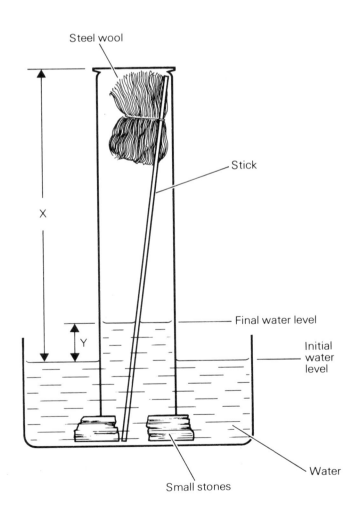

Steel wool

X

Stick

Y

Final water level

Initial
water
level

Water

Small stones

9

(2) Analysing dust from the air **

When a beam of sunlight or torchlight passes through air in a darkened room, you can see minute particles of dust dancing and moving about. This dust builds up in ventilators, corners of rooms, vacuum cleaners and below radiator shelves. Let us analyse it.

What you will need
A Petri dish or similar container
A hand lens or microscope
A crucible or silica dish
A balance
A bunsen burner, pipe clay triangle and tripod stand

What to do
Place a small amount of dust on the lid of a Petri dish. Look at it through a hand lens, or a microscope if you have one.

Sort out the make-up of the dust. Human hair is easy to recognize, and so is wool. Cotton fibres are more difficult but they are twisted and can be identified by comparing them with cotton wool or lint. You might also find plant seeds, pieces of dried leaf, pollen grains, soil particles and fragments of skin. Make a rough count of the proportions of hair, wool, cotton and seeds.

The quantity of minerals in the dust can be worked out by burning off the organic matter. To do this weigh out a small sample, about 0.25 gramme, in a crucible or silica basin. Remember, a large volume of dust weighs very little. Burn the sample over a gas flame, an electric heater or in a muffle furnace. Re-weigh the dish when it

has cooled. The percentage of mineral material can be calculated as in this example. (The figures are taken from an actual experiment):

Weight of dust before burning	= 0.3600 gramme
Weight of dust after burning	= 0.3284 gramme
Loss of weight by burning	= 0.0316 gramme
% of mineral material	= $\dfrac{\text{loss of weight} \times 100}{\text{original weight}}$
	= $\dfrac{0.0316 \times 100}{0.3600}$
	= 8.7%

The organic content therefore, must be 91.3% (i.e. 100% − 8.7%). The dust used was taken from a ventilator in a college building. Dust from other sources will have different proportions of mineral and organic matter. For example, in an experiment, the dust from a domestic vacuum cleaner contained, on average, 19.0% mineral matter.

To stop dust and other debris entering the lungs, mammals (including humans) have a special bone at the back of the nose. Air must pass over this before it can enter the lungs. The membrane covering the bone, plus the hairs at the entrance to the nose, act as filters. The air entering the lungs is thus relatively clean. However, when the percentage of dust in their air is high, much of it enters the lungs.

(3) Measuring the soot content of the atmosphere **

The soot content of the air can be measured in two ways: first by letting soot settle out of the air onto a white tile, and second by passing a large volume of air through a filter paper.

Method 1

What you will need
One matt or unglazed white tile

What to do
The first method, although the simplest, is the least accurate. This is because the deposition of small particles onto a flat surface depends on many things. To get the best results, place the tile outdoors in a sheltered corner. Examine it daily and count 'the number of soot specks deposited during each time interval. After each examination the tile should be wiped clean. Try to find out on which days of the week there are the least deposits, and why this is the case.

This first experiment does not measure exactly how much soot is present, but useful comparisons may be made if several sites are used at the same time. In our experiments four sites were used.

Tall chimney stacks at Corby in Northamptonshire.

These were: inside a closed drawer, on a shelf in a study bedroom, on a window sill 10 m above the ground, and on the surface of the soil itself. The number of soot particles counted on each tile is given below. The table shows the difference between the rate of soot deposition at the four sites:

Table 2. The deposition of soot on tiles

Site	Day 1 (rain)	Day 2 (dry)	Day 3 (rain)	Day 4 (rain)	Day 5 (damp)	Day 6 (breezy)	Day 7 (wind and rain)	Totals
Inside drawer	1	8	7	1	2	7	12	38
On shelf in study	10	47	42	32	39	45	58	273
On window sill	20	108	102	99	78	112	101	620
On the ground	33	123	127	90	103	115	123	714
Daily totals	64	286	278	222	221	277	294	1642

Left Diagram 2(a) Graph to show the deposition of soot on white tiles.

Below Diagram 2(b) Graph to show the deposition of soot at four locations.

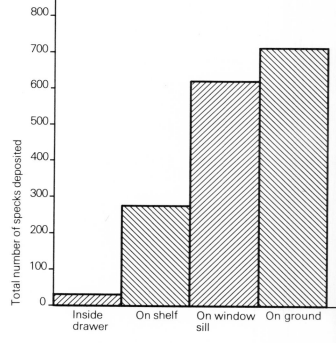

Total number of specks deposited — Number of days

Total number of specks deposited — Location

Method 2(a)

What you will need

A round piece of stiff card 10 cm in diameter
A pair of compasses
A sharp knife or scalpel
A large funnel, either glass or plastic
Some plasticine, some glue and a length of thin wire
Several metres of rubber tubing to fit the stem of the funnel
A water-operated vacuum pump
A balance
A hand lens
A flow meter or a 1 m glass tube 1 cm in diameter
A light meter
Some 10 cm diameter filter papers

What to do

In the middle of the piece of card cut out a round hole 5 cm in diameter. Glue the 10 cm filter paper to the card disc. The card and filter paper should now be fastened inside the funnel with plasticine. Hang the funnel upside down outside the window and support it by a simple frame of thin wire. Attach the rubber tubing to the funnel at one end, and to the water pump at the other (diagram 3). Switch on the pump and draw air through the paper for at least twenty-four hours. Then remove the card and paper from the funnel and examine it with a hand lens. Lots of scraps of soot and other debris should be clearly visible.

To get accurate figures of the amount of soot in a cubic metre of air, the filter paper and card should be dried at 105°C and then weighed before the start of the experiment. The rate of flow of air through the filter paper must also be measured. This can be done using a flow meter or a soap bubble meter. A soap bubble meter can be made by turning a 50 cm³ burette upside down in a trough of soap solution. As air is drawn through the burette a soap bubble travels along the length of the tube. The time taken to travel from the 0.0 cm³ to the 50 cm³ marks is measured, and hence the rate of flow in cm³ per second can be worked out (diagram 4a).

At the end of the experiment the filter paper and card should be re-weighed after drying once more at 105°C. The increase of weight is the amount of soot from the total amount of air drawn through; this volume is worked out from the rate of flow and the time taken to finish the experiment.

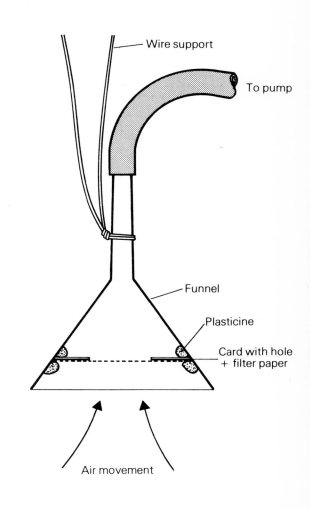

Diagram 3 Apparatus for filtering soot from the air.

13

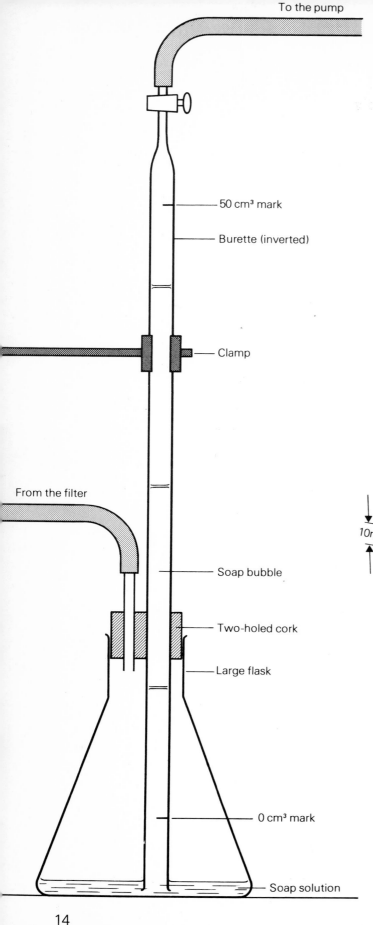

To the pump

50 cm³ mark

Burette (inverted)

Clamp

From the filter

Soap bubble

Two-holed cork

Large flask

0 cm³ mark

Soap solution

What the results can show

In one experiment 0.0131 gramme of soot was deposited in forty-eight hours. The rate of flow was 3,000 cm³ per minute. Therefore, 8.64 m³ of air had passed through the filter paper. Therefore, one cubic metre of air contains 0.0015 gramme of soot.

Put another way, a column of air 10 m high covering an area of ground of 100 metres square contains 150 gramme of soot. Assuming all the soot is deposited this is approximately 28 kg of soot deposited in a year on a piece of ground 100 metres square (diagram 4b).

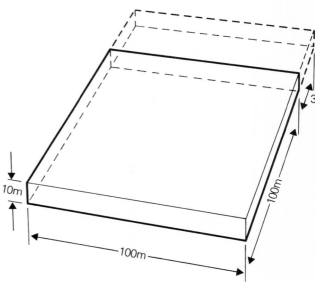

10m

100m

100m

3(

Above Diagram 4(b) This block of air 10 × 100 × 100 m contains 150 gramme of soot. (The dotted area represents the extra ground area required to make a football pitch).

Left Diagram 4(a) A soap bubble flow meter.

Method 2 (b)

To avoid weighing the soot, the strength of the soot deposit may be found by reflecting light from the surface of the soot-covered filter paper. A photographer's light meter may be used. If this is marked in Lux, standard light measurements may be taken. All of the filter papers will need to be exactly the same if accurate measurements are to be made.

What the results can show

These results give some idea of the efficiency of this method:

Table 3. Light reflected from various filter papers

Type of filter paper	Light reflected (Lux)
Clean filter paper	15
Unglazed black paper	5
Air passed for 48 hours	13
Air passed for 67 hours	12
Air passed for 74 hours	14 (new weather conditions — windy)

Soot particles and tar deposited on a glass slide placed inside a chimney (magnified 120 times).

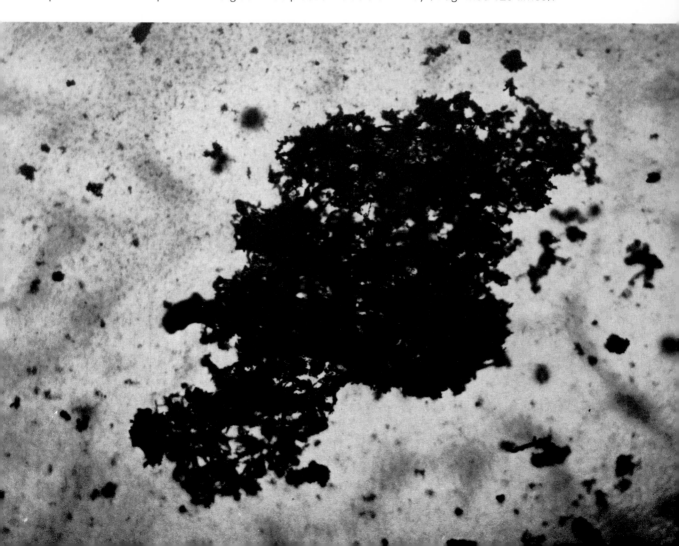

(4) Identifying the components of atmospheric dust ****

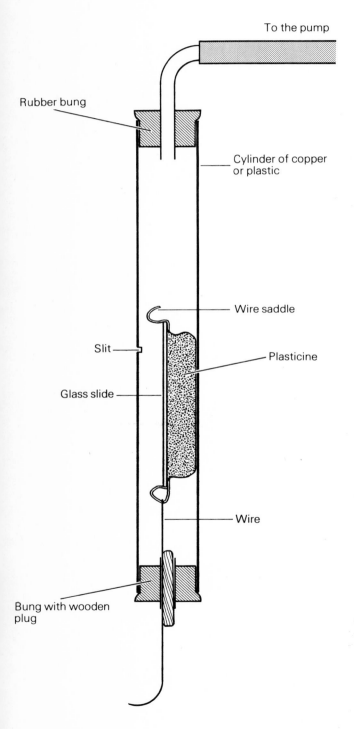

To the pump

Rubber bung

Cylinder of copper or plastic

Wire saddle

Slit

Plasticine

Glass slide

Wire

Bung with wooden plug

In previous experiments, soot and other materials have been taken out of the air and examined. In this one we will try to work out what all of the materials in the air actually are.

What you will need
One cylinder of cardboard, plastic or metal (an off-cut from a 25 mm copper water pipe is good for this)
A rubber bung fitted with a glass tube (to fit the cylinder)
A length of rubber tubing to fit the glass tube
A microscope
Some glass microscope slides
A water pump
An alarm clock
Some thin copper wire
A little plasticine and some Vaseline

Diagram 5 The slit sampler. **Left** — longitudinal section, **below** — cross-section

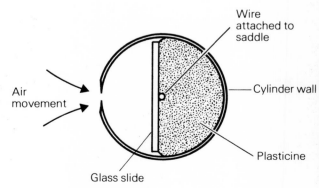

Wire attached to saddle

Air movement

Cylinder wall

Plasticine

Glass slide

What to do

In one side of the cylinder, make a small slit 2 mm by 3 mm long, about 8 cm from one end. Shape the plasticine into a segment 8 cm long that will just fit inside the cylinder (diagram 5). Using copper wire, make a small saddle which will fit and hold a slide. Now smear Vaseline on the slide. Attach the saddle with a long length of wire fixed to one end. Stick the slide to the plasticine with the Vaseline side facing upwards. Push the plasticine and slide into the cylinder so that the Vaseline surface of the slide is opposite the slit. Place one of the rubber bungs fitted with a glass tube in the top of the cylinder; attach the rubber tubing from the water pump. Pass the wire from the saddle through the glass tube in the second bung, and slide into the lower end of the cylinder. Fill the hole on the lower bung, where the wire comes out, with a wooden plug. Hang the tube outside the room window and turn on the water pump.

The apparatus may be left on for twenty-four hours. Then, remove the slide from the cylinder and examine it

Diagram 6 Low power drawing from the slide in a slit sampler.

under a microscope. With practice, you will be able to identify all the components of the dust. Soot specks are easy to see; sand grains from bricks and mortar look like regular "see-through" cubes. Spores of fungus and pollen grains are harder to identify but with practice each type may be recognized.

We have known for many years that the types of spores in the air change from day to day because spore release changes with the weather and with the time of day or night. To chart these changes, the slide should be moved downward in the slit sampler. To do this pull the wire joined on to the slide saddle at regular intervals of time. Any bands on the slide will then show what spores are present at different times of the day.

For the technically minded. The amounts of different materials collected on the slide can be estimated by counting the different kinds of particles in one area as seen through a microscope and recording the results for each kind of particle. If this is repeated 20-30 times, after moving the microscope to another part of the slide, then an accurate average can be found of the numbers of each kind of particle present.

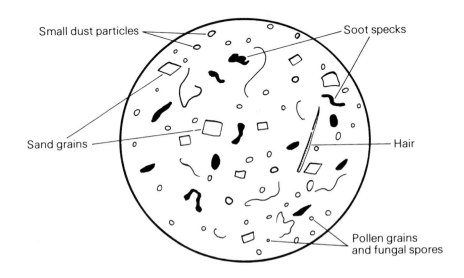

Small dust particles — Soot specks — Sand grains — Hair — Pollen grains and fungal spores

(5) Weighing the deposition of soot on plant leaves **

In towns and cities the privet hedge plant is a popular border for gardens. This bush has broad evergreen leaves that survive for two to three years. In towns the leaves often have a layer of soot on their upper surfaces. By taking a two-year old leaf, the amount of soot deposited during the life of the leaf can be measured. Three ways of doing this are described. One uses a "home made" straw balance; the second a more complicated and expensive chemical balance. A third involves comparisons of leaves from various sites.

Method 1: Using a straw balance
What you will need
A paper or plastic straw
A small screw of the same diameter as the straw
Some graph paper
A small, stiff-bristled paint brush
A metal channel
A fine needle
Some microscope slides
A wooden block

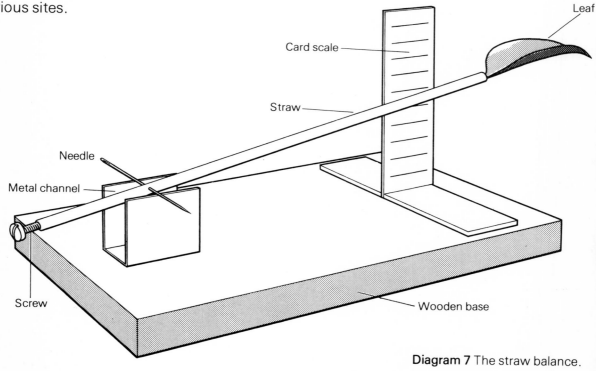

Diagram 7 The straw balance.

What to do

Stick a needle through the straw, about 3 cm from one end. Cut away part of the straw at the long arm end to make a little scoop; this is where the things to be weighed are placed. Balance the needle axle on the metal channel and insert the small screw as a counterweight in the short arm of the straw. When the straw is balanced, use an upright piece of card to make a scale for the long end of the straw. This may be held in place by sandwiching it between two microscope slides. For weights, use squares of graph paper. if the weight is needed in gramme then 100 sheets of graph paper may be weighed on a balance and the weight of one square calculated.

Now take a dirty leaf. Place it in the long end of the straw and stick it in position with sticky tape. Balance by adjusting the screw. Now brush off the dirt with the stiff-bristled brush. The long arm of the balance will move upwards. Re-adjust to the horizontal by adding graph paper "weights".

Using graph paper, work out the total surface area of the leaf. This can be done by drawing an outline around the leaf onto a sheet of graph paper and then counting the number of squares inside it. Knowing the weight of dirt on the leaf, and the area of its surface, the amount of dirt per unit area may be calculated.

The following results were reached using leaves from a roadside location:

Method 2: Using a chemical balance

What you will need
Some privet leaves
An evaporating dish
Some distilled water
A chemical balance
One pair of fine-pointed forceps
A small, stiff-bristled paint brush

What to do

Take five or more privet leaves of a regular shape. Make sure they have a layer of soot on them. Regular-shaped leaves are best because it is easier to measure their area. Pour about 10 cm³ of distilled water into a dried and weighed evaporating dish (or any clean glass dish). Using fine-pointed forceps to grip the leaf petiole (stalk) place each leaf, in turn, in the water. Brush the upper surfaces with a small, stiff-bristled paint brush. At the end of the cleaning process for each leaf, hold them above the water and rinse with a jet of clean water from a wash bottle.

Cover the dish with a clean sheet of paper. Place it in a warm cupboard, over a radiator or in a drying oven, to let the water evaporate. While the dish is drying, blot the leaves and measure their area using the squared-paper method.

Re-weigh the dried dish and soot. The difference between the weight obtained and the original weight is the amount of soot on the leaves.

The results shown on next page were reached in one experiment.

Table 4. Weight of soot on roadside leaves

Weight of 100 sheets of graph paper   = 504.6 gramme
Weight of 1 sheet of graph paper   = 5.046 gramme
1 sheet of graph paper   = 600 × 1cm squares
1 cm² of graph paper weighs 5.046 ÷ 600 gramme
1 cm² of graph paper weighs 0.0084 gramme
¼ cm² of graph paper weighs 0.0021 gramme

Leaf	No of ¼ cm² used to re-balance after the leaf was cleaned	Weight equivalent of ¼ cm²	Surface area cm²	Weight per m² per annum
1	1 ¾	0.0037	9.0	4.1 gramme
2	1	0.0021	8.0	2.5 gramme
3	1 ¼	0.0027	8.3	3.3 gramme

Five 2-year-old privet leaves from an industrial area in England, having a total surface area of 40.7 cm², carried 0.0155 gramme of soot.

What the results showed
In the area where the leaf was growing, 0.0155 gramme of soot fell per 20.35 cm² over two years.

In the area where the leaf was growing 0.0155 × 10,000/20.35 gramme of soot fell per square metre over two years.

In the area where the leaf was growing 7.6/2 gramme of soot fell per square metre per year.

In the area where the leaf was growing 3.8 × 10 kg of soot fell annually, per 100 metres square. That is, 38 kg per 100 metres square.

This compares very well with the figure of 28 kilograms obtained by experiment 3 (method 2a).

Method 3
What you will need
Some privet leaves
An evaporating dish
Some distilled water
A stiff-bristled brush
One pair of fine-pointed forceps
Some filter paper

What to do
First wash the leaves using the process described in Method 2. Once the leaves have been washed, pour the suspension of soot and water through the filter paper and dry. Record the amount of soot.

What the results show
This method does not give you an exact quantity because no weighing is taking place. However, by keeping filter papers from different sites, comparisons between areas may be made..

Heavy atmospheric pollution from a factory chimney.

(6) Measuring the shade effect of soot on plant leaves ****

What you will need

A cardboard tube approximately 5 cm in diameter and 25 cm long
A light meter
A desk lamp and 75 watt bulb
Some black card or stiff black paper
A roll of sticky tape
A retort stand and clamp
A safety razor blade
A cork borer
A number of clean and dirty leaves from a roadside or industrial area

What to do

Find the mid-point of the length of tube. At this spot, slice through the tube three-quarters of the way through. The uncut portion of the tube acts as a hinge.

Cut a round disc of black card the same diameter as the tube. Cut a 1.5 cm diameter, round hole at its centre. Fix the disc in position with sticky tape so that it closes one end of the tube at a point near to the hinge (see diagram 8).

In one of the original tube ends place a light meter so that the light-receiving surface points into the tube. (In our experiments we used a light meter with a circular light-receiving area which just fitted inside the tube.)

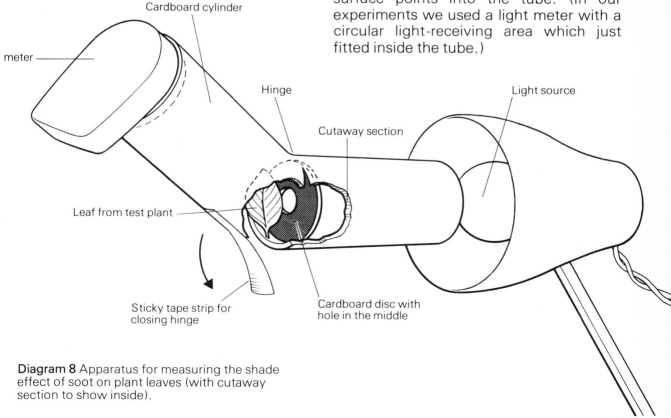

Diagram 8 Apparatus for measuring the shade effect of soot on plant leaves (with cutaway section to show inside).

Close the two halves of the tube, seal it with a strip of sticky tape and fix in the retort clamp horizontally. Direct the light from the lamp along the tube and take the reading on the light meter. Open the tube. Place one of the dirty leaves behind the black disc, fixing it in place with sticky tape. Record the amount of light passing through the leaf.

Repeat, using either a clean or a washed leaf from the same site.

In practice several leaves of each type, clean, dirty and dirty but cleaned, should be tested for their ability to transmit light.

Table 5 summarizes the results of a series of experiments:

Table 5. Measuring the shade effect of soot

Clean leaf	Dirty leaf	Cleaned dirty leaf
70 Lux	50 Lux	60 Lux
70 ,,	50 ,,	80 ,,
80 ,,	50 ,,	90 ,,
50 ,,	40 ,,	90 ,,
40 ,,	30 ,,	90 ,,
50 ,,	30 ,,	70 ,,
150 ,,	110 ,,	—
130 ,,	50 ,,	—
175 ,,	80 ,,	—
100 ,,	50 ,,	—
90 ,,	60 ,,	—
150 ,,	80 ,,	—
120 ,,	75 ,,	—

Points to remember

(i) The shading effect of soot, etc, is shown by a marked decrease in the amount of light passing through the leaf. This will affect the rate of photosynthesis (see diagram 9).

(ii) If the light reaching the leaf is very bright the shade effect will make little difference to the rate of photosynthesis, because there will be a surplus of light falling on the leaf. It is known from other experiments that plants cannot use *all* the light reaching their leaves.

(iii) The shade effect will have the greatest effect on plants growing in areas where the atmosphere is full of gases and low ceiling clouds. In such areas, plants will have small annual growth rates. Poor vegetation is often a feature of older city centres.

Diagram 9 How plants feed — clean leaves are essential for the process of photosynthesis.

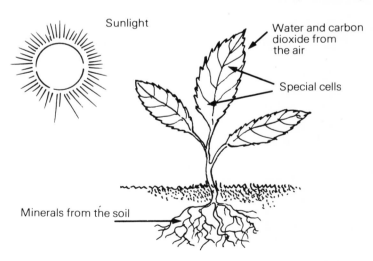

Sunlight

Water and carbon dioxide from the air

Special cells

Minerals from the soil

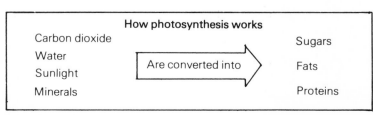

How photosynthesis works		
Carbon dioxide		Sugars
Water	Are converted into	Fats
Sunlight		
Minerals		Proteins

Above Fumes emitted by a tin smelter and . . .

below the effects these fumes have upon trees in the area.

(7) Measuring the rate of photosynthesis in leaves contaminated with soot *****

In the previous experiment (6) the shade effect of soot was measured by the in-crease in the amount of light passing through leaves after they had been cleaned. By comparing the results obtained from (6) with those from this experiment we can correlate the shade effect and the reduced rate of photosynthesis.

What you will need
M/1000 solution of sodium hydrogen carbonate indicator solution (0.084g/litre + cresol red)
Three very clean boiling tubes 3 cm in diameter, fitted with rubber bungs
A glass aquarium
A thermometer
A 100 watt lamp
A No 12 cork borer
A boiling tube rack
Some cotton wool
A selection of clean and dirty leaves

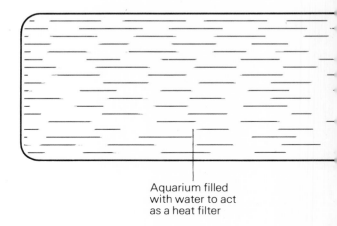

Aquarium filled with water to act as a heat filter

Boiling tubes + leaf discs

Diagram 10 Apparatus for measuring the rate of photosynthesis of leaves contaminated with soot.

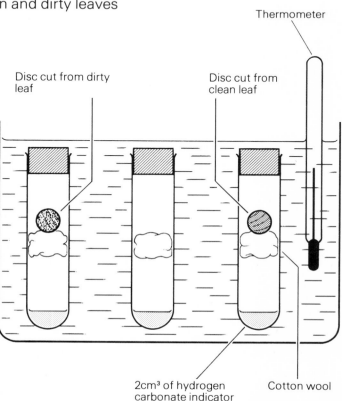

Thermometer

Disc cut from dirty leaf

Disc cut from clean leaf

2cm³ of hydrogen carbonate indicator

Cotton wool

What to do

The indicator solution should be exposed to air from outside the laboratory when it will have a cherry red colour. Place 2 cm³ of this solution in each of the boiling tubes. Push in a loose wad of cotton wool halfway down the tube. On top of the cotton in one tube place a round disc cut from the dirty leaf. In the second tube place a disc cut from the clean leaf. The discs should be from leaves of the same species, similar age, and cut from the same position on the leaf. The third tube is left without a leaf disc to act as a control.

Place the tubes in the rack. Put the rack behind an aquarium tank, about 15 cm away from the glass. Place a thermometer in the aquarium and switch on the light. The water acts as a heat filter so that the rays of light are not hot when they reach the leaf (see diagram 10).

Record the colour of the indicator solution every five minutes. As the carbon dioxide is removed from the air in the boiling tube, by the photosynthetic activity of the leaf discs, the colour of the indicator will change from red to deep purple.

This table summarizes some of the results obtained by this method:

Table 6. The effect of dirt on the rate of photosynthesis

Time in minutes	Control tube		Clean leaf		Soot covered leaf	
	Experiment 1	Experiment 2	Experiment 1	Experiment 2	Experiment 1	Experiment 2
5	No change	No change	No change	No change	No change	No change
10	No change	No change	No change	No change	No change	No change
15	No change	No change	No change	No change	No change	No change
20	No change	No change	No change	No change	No change	No change
25	No change	No change	No change	No change	No change	No change
30	No change	No change	No change	No change	No change	No change
35	No change	No change	No change	No change	No change	No change
40	No change	No change	No change	No change	No change	No change
45	No change	No change	No change	No change	No change	No change
50	No change	No change	Deep red	Deep red	No change	No change
55	No change	No change	Deep red	Deep red	No change	No change
60	No change	No change	Light purple	Red	No change	No change
65	No change	No change	Light purple	Light purple	No change	No change
70	No change	No change	Light purple	Light purple	Deepening red	Deepening red
75	No change	No change	Light purple	Light purple	Deep red	Deep red
80	No change	No change	Light purple	Light purple	Light purple	Light purple

What the results show

(i) Leaves with upper surfaces covered with soot have a lower   of photosynthetic activity than clean leaves.

(ii) The reduced rate of photosynthetic activity, as measured by the time taken for the indicator to change, was almost constant. (Other results not recorded confirmed this.) The time difference was 15 minutes.

(iii) The results confirm the shading effect of grime. They suggest that plants growing in polluted areas will have slower growth rates than plants growing in cleaner areas.

(8) Measuring the amount of sulphur dioxide in the air

When fuels such as coal, coke, fuel oil and wood are burned, large amounts of gas rise into the atmosphere. The main gases are carbon dioxide and carbon monoxide but there is also water vapour present.

However, almost all fuels contain sulphur compounds. The quantity of sulphur compounds varies from 0.01% to 4%. When the fuel is burned the sulphur compounds are changed in to sulphur dioxide. This gas is extremely unpleasant, and irritates the lungs even in small amounts. Fortunately it dissolves easily in water and will be washed out of the atmosphere by rain. The product formed when it dissolves is a weak acid — sulphurous acid. In industrial areas sulphurous acid seeps into the soil making

it useless for growing plants. Farmers may have to add large amounts of lime to the soil to fight this problem.

Following recent legislation, the quantity of smoke in the atmosphere has been greatly reduced in many areas. But the atmosphere is still far from pure, and the amount of sulphur dioxide in the air has continued to rise until recently (diagram 11). When smokeless fuels are burned there is little smoke, but the quantity of sulphur dioxide produced is still high. Apart from its irritant effect on the nose, eyes and throat, sulphur dioxide affects the growth of plant and animal life. In particular, lichens are very sensitive to the gas and now very few lichens are to be found in our big cities.

It is impossible, with our simple apparatus, to make an exact measure of sulphur dioxide in air. But since this gas is soluble in water, we can estimate how much there is around by collecting rainwater and measuring its pH. The following experiments give some indication of the presence of this acid gas in air.

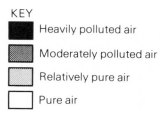

KEY

■ Heavily polluted air

▨ Moderately polluted air

▧ Relatively pure air

□ Pure air

Diagram 11 Map to show the amount of sulphur dioxide in the air over the UK in 1972 (from *The Sunday Times Magazine,* January 28th, 1973).

The muggy, smoke-filled air over the town of Manchester in England.

Method 1*
What you will need
A large funnel, made of glass or plastic
A glass jar
A book of pH test papers

What to do
Collect rainwater from downspouts, freshwater ponds and garden water-butts. Dip in a strip of pH paper. Match the colour obtained with the colour chart issued with the pH papers.

In our experiments we found that rainwater had a pH of 5.4. In other words it was quite acid. This shows that there was a fairly high sulphur dioxide content in the air.

Method 2*
What you will need
A trowel
Some plasticine
A few glass containers
Some pH paper (or better still, a pH meter)

What to do
Many trees have rainwater channels along their branches and down the side of their trunks. After falling on the leaves, the rain drips off onto the branches and runs towards the earth along these channels. These "run-off" channels also continue away from the base of the trunk and into the soil. It is easy to plot the channels in the soil noting by the absence of plant life.

Find a suitable tree and make a small dam out of plasticine in a "run-off" channel when the bark is dry. When it is raining, water collects behind the dam. It is quite simple to test the pH with pH paper. Or, collect a sample in a jar and measure the pH using a pH meter.

Another way of getting a sample is to scrape some of the bark from the tree, or remove some of the soil from the "run-off" channel. Soak the bark or soil in distilled water for a few minutes. After soaking, filter and test the filtrate with pH paper or a pH meter, as before.

To get a "control" measurement take some ordinary soil from an area away from the "run-off" channel and repeat the experiment. This soil may also have an acidic pH due to the general rainfall, but it is still worth comparing the results.

We got the following results using a pH meter:

Table 7. pH of run-off water as an indication of sulphur dioxide in air

Position	Type of tree:		
	Beech	Wych elm	Elm
Run-off channel	3.0	3.1	3.4
Ordinary bark	3.3	3.3	3.4
Soil from channel			3.4
Soil away from channel			4.4

Ordinary rainwater	5.4
Rainwater from roof channel	4.0
Tap water	7.0
Distilled water	7.0

Method 3 ***
What you will need
A tall jam jar
A jar of known volume (preferably 1 litre)
An aquarium pump, airline tubing and diffuser stone
A dilute solution of hydrogen peroxide from the chemist (used for bleaching hair)
Some barium chloride solution (1 gramme dissolved in 500 cm³ of water. Take care as barium salts are very poisonous)
Some vinegar or other dilute acid (but *not* sulphuric acid)

What to do
Connect up the pump to the diffuser stone with the airline tubing. Plug the pump to the power supply. Dip the diffuser stone in 250 cm³ of hydrogen peroxide in a tall jam jar. Blow air into the solution for 4 days.

After this time acidify 50 cm³ of the solution with a little vinegar, or other acid and add 50 cm³ of barium chloride solution. A white precipitate will form. The white precipitate may form after less than four days or more than four days depending on the amount of sulphur dioxide in the air in your area. If the white precipitate can just be seen then the solution contains about 12 milligramme of sulphur dioxide. As the precipitate gets denser so the number of milligrammes increases.

In order to work out how many milligrammes were in the original air the amount pumped out over 4 days must be known. This can be calculated by finding out how long it takes to fill the 1 litre jar with gas from the diffusor stone. The example result will put you in the picture.

What the example result showed
In a test, the following result was obtained:
1 litre of gas was pumped into the jar in 3 minutes. So, 20 litres would be pumped out by the pump in 1 hour and 1,920 litres would be pumped out in 4 days (96 hours x 20 litres). This amount is approximately 2,000 litres or 2 cubic metres. This amount of air produced a white precipitate which could just be seen, so the solution contained 12 milligramme of sulphur dioxide. 12 milligramme of sulphur dioxide was present in 2 cubic metres of air. So, 6 milligramme of sulphur dioxide was present in 1 cubic metre of air. This is 6 parts per million. (Remember that 1 milligramme in 1 cubic metre is 1 part per million. For a solution, 1 milligramme in 1 litre is 1 part per million.) There are 6 parts per million of sulphur dioxide in the air tested.

(9) The biological detection of sulphur dioxide in the atmosphere ***

Another way of estimating the amount of sulphur dioxide in the air is by looking for indicator plants sensitive to this gas. Lichens are a good example of such a group of plants. Lichens are made up of a fungus and a green plant (algae). They live very close together and grow on stones, trees, rooftops and on buildings where the level of pollution is very low.

There are three main types of lichens:

(i) *The shrubby or fruticose type:* These look like leafless miniature bushes, from two to five cm high. They are grey-green in colour, and are highly sensitive to pollution. They grow only in remote areas such as Dartmoor, parts of the Lake District and the mountains of Wales.

Below Fruticose lichen.

(ii) *Leafy or foliose type:* These have the appearance of flat leafy lobed plants growing on stones and trees. Again their colour is grey-green. They

Above Foliose lichen.

can put up with more pollution than the fruticose type. They may be found in the middle regions between towns and remote upland areas.

(iii) *The crustaceous lichens:* These lichens are so named because they grow tightly attached to the surface below them. Some of these types are brightly coloured green, orange and white. They look like paint splashes. Such lichens are the most tolerant of pollution and may be found on the outskirts of towns growing on asbestos or clay tile roofs. Those growing on asbestos can live with more pollution because cement asbestos can neutralize the acid rainwater.

Below Crustaceous lichen.

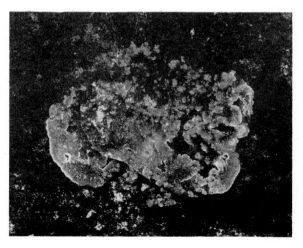

What you will need

An Ordnance Survey Map of the local area

A compass

A few sheets of 1 cm squared paper

What to do

Use the map to work out where the town or city centre is, and visit it. Look at the buildings, especially church yards, for traces of lichens on walls, headstones, and so on. Normally city centres are empty of lichens.

Now mark lines on the map, spreading outwards from the city centre. Walk outwards from the city centre looking for the various classes of lichens. If the town is large, distances between inspection sites may be at least a kilometre. As the distances from the centre become greater, look more carefully, especially at the bases of old walls and older headstones.

Diagram 12 Wind direction and the distribution of lichens around a city centre.

Mark on the map, or make sketches of the map on squared paper, showing all the places where lichens are found. Record the findings as (a) rare, (b) frequent, or (c) very common. Or make your own system of records. A system based on the number of lichens per unit area (e.g. per square km) is best, but if you have made only a very few or very many finds it is a hard system to operate.

If you have the time and a means of transport, you may be able to draw an "area locational distribution map". This will bear in mind the prevailing wind direction as well as the heaviness of the pollution. We would expect the distribution to be broadly the same shape as that shown in diagram 12. The actual area of the lichen desert will grow as the amount of pollution at the city centre increases.

What the results show

The results show the extent of sulphur dioxide pollution sufficient to cause a lichen desert.

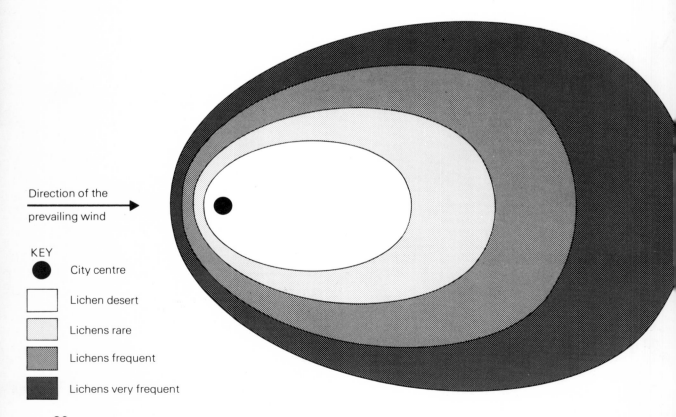

Direction of the prevailing wind

KEY

● City centre

☐ Lichen desert

▦ Lichens rare

▨ Lichens frequent

■ Lichens very frequent

(10) Trapping fungal spores and pollen grains from the atmosphere *****

Above Parasol mushrooms. Like other types of fungi, these reproduce through spores or primitive seeds.

What you will need
A slit sampler similar to that used in experiment (4)
Some microscope slides and a microscope
A little Vaseline

What to do
Make and use the slit sampler as in experiment (4). The best time of year to trap pollen grains and fungal spores is in summer and early autumn when there are large amounts floating in the air. At the end of each experimental period examine the slides through a good microscope. You can identify individual spores and grains either by using standard text books, or by making up reference slides of pollen and spores from known grasses, trees and fungi (a highly-skilled tech—nique).

If the slide can be moved regularly over a period of hours, the time of day when particular species shed their pollen or spores may be determined. By counting the frequency of particular types of pollen or spore, the make-up of spores in the air may be discovered.

II Fuels as a source of pollution

Introduction

Millions of tonnes of fossil (or carbon-iferous) fuels are burned every year for heat and power. Coal, gas and oil are all used to heat homes, factories and offices, to provide hot water and steam, and to provide power for cars, lorries and trains. Power stations use vast amounts to make high-pressure steam, to drive the turbines which generate electricity. Air-craft burn aviation fuel in the upper

atmosphere; ships burn fuel oil at sea. All of this fuel is drawn from a fast-dwindling stock which has been in the earth for millions of years.

Decomposing vegetable matter gives off methane, or natural gas (see experiment 20). Rocks sometimes trap this natural gas. Coal is the squashed remains of the fern forests of the carboniferous age. Oil comes from the remains of marine life feeding on plankton. Oil may also come from the tarry oils made by plants.

Because they come from carbon dioxide and water all carboniferous fuels contain the elements carbon and hydrogen. So when carboniferous fuels burn they mainly produce carbon dioxide and water vapour. Other elements that are present in plants, and so in carboniferous fuels, are sulphur and nitrogen. These elements produce unpleasant gases such as sulphur dioxide, hydrogen sulphide, hydrogen cyanide and ammonia, when the fuel is burned. All of these gases escape into the atmosphere. The fuels also contain stone, sand and other inorganic matter. Most of this is left behind as ash, but some escapes into the air as smoke. In fact, smoke is a suspension of very small particles of fuel and

On facing page Pittsburgh (USA) in the 1940s before anti-pollution laws were brought into force, and **below** Pittsburgh in 1970.

grit which have not been burned.

One of the main pollutants is petrol and diesel oil used by cars, lorries and trains. Both fuels are a complex mixture of hydrocarbons (chemicals made up of carbon and hydrogen only). In car and lorry engines the hydrocarbons are mixed with air and burned to make exhaust gases, heat and power. If the hydrocarbons burn well the exhaust gases will be carbon dioxide and steam. But a car sometimes idles, sometimes accelerates and sometimes decelerates, so that the hydrocarbons cannot burn well all the time. This means that the exhaust gas will contain carbon monoxide as well as carbon dioxide and steam.

Carbon monoxide is a very poisonous gas and one part in 600 parts of air is enough to kill an adult human being. The gas normally spreads out quickly enough so that it does not become a danger but in heavy traffic the amount could build up as it would in a closed garage.

Most petrols contain other chemicals called additives. High octane petrol contains up to 1 cm³ of tetraethyl lead (TEL) per litre of fuel. This is burned by the vehicle to make lead oxide. This becomes deposited as a white powder on the sparking plugs stopping them from working. So ethyl bromide, another additive, is added with the tetraethyl lead. On burning, the lead is changed to a gas, lead bromide, which goes into the atmosphere along the exhaust pipe together with all the other exhaust gases.

Unfortunately lead compounds are poisonous and can cause permanent brain damage and general restlessness. Some countries are thinking of reducing or eliminating the lead content of petrols.

All fuels contain sulphur compounds which burn to give a gas called sulphur dioxide. This gas is poisonous in large amounts and irritates the lungs even in small amounts. There are strict laws to make sure that fuels do not contain too much sulphur but it is impossible to remove it completely. So much fuel is burned today that even with strict laws the amount of sulphur dioxide produced is many millions of tonnes per year.

The sparking plugs of a petrol engine produce a spark to ignite the mixture of air and petrol which then drives the pistons. The spark also causes minute amounts of nitrogen and oxygen to combine to form a brown gas called nitrogen dioxide. This is again an irritant and poisonous in large amounts. Fortunately, it is very soluble in water and does not seem to be a problem in the wet countries of Western Europe. In places like Los Angeles or Tokyo, however, the nitrogen dioxide and other fumes from cars combine in the energy of sunlight to form a very obnoxious smog which irritates the eyes and even kills plants. This state of affairs is very dangerous and new laws have been passed requiring cars to have special exhaust systems which will absorb nitrogen dioxide.

Table 8. The approximate composition of motor vehicle exhaust gases

Component	Petrol engine	Diesel engine
Carbon dioxide	8.4 %	1.8 %
Carbon monoxide	3.3 %	0.03%
Oxygen	6.4 %	18.83 %
Methane	0.18 %	0.03 %
Hydrogen	1.1 %	—
Nitrogen	80.5 %	79.9 %
Nitrogen oxides	600 ppm	400 ppm
Sulphur dioxide	60 ppm	200 ppm
Aldehydes	40 ppm	20 ppm
Lead bromide	20 ppm	none

(11) Measuring the amount of carbon dioxide, carbon monoxide, oxygen and nitrogen in exhaust gases ****

What you will need
Two gas syringes of 100 cm³ capacity
Two plastic (or glass) T-pieces
Two boiling tubes
Two rubber bungs with two holes in them to fit the boiling tubes
Two aquarium bubblers
Some glass tubing to fit the rubber bungs
Some rubber tubing to fit the glass tubing
A large plastic bag

A little strong caustic potash solution
Some strong alkaline pyrogallol solution
Some ammoniacal silver nitrate solution or ammoniacal copper (I) chloride* solution

*Warning: These are dangerous solutions: Use rubber gloves and wash off any spills with water

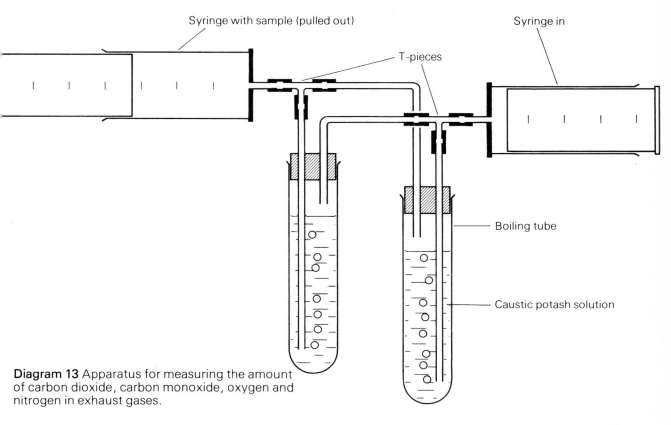

Syringe with sample (pulled out)

Syringe in

T-pieces

Boiling tube

Caustic potash solution

Diagram 13 Apparatus for measuring the amount of carbon dioxide, carbon monoxide, oxygen and nitrogen in exhaust gases.

What to do

Build up the apparatus as shown in diagram 13. Collect a sample of exhaust gas from a car by placing the open end of the plastic bag over the end of the exhaust pipe. If the pipe is hot, wrap asbestos cloth over the end first, or use household gloves. Run the engine for about two minutes until the plastic bag is full. Seal up the bag with a piece of string and take it back to the laboratory.

(i) *Measuring the amount of carbon dioxide:* In the laboratory take a sample of gas from the bag with one of the syringes, and re-seal with sticky tape. Adjust the amount of the sample in the syringe to 100 m³. Connect the syringe to the apparatus containing caustic potash, taking care that the other syringe is fully closed.

Now gently push in the plunger of the syringe with exhaust gas in it. At the same time the plunger of the empty syringe should be gently pulled out. When all the gas has bubbled through the caustic potash solution, reverse the flow of gas by pulling out one syringe and pushing in the other. In this way the gas is passed through the caustic potash solution several times. As this happens the volume of gas slowly drops. When no further drop in volume occurs, read off the exact reduction in volume on the scale marked on the syringes. Do the experiment two or three times and obtain an average result. Now, caustic potash is able to absorb carbon dioxide. It therefore follows that the reduction in volume of exhaust gas must equal the percentage volume of carbon dioxide present in the original exhaust gas sample.

(ii) *Measuring the amount of carbon monoxide:* Now replace the boiling tubes of caustic potash with boiling tubes of ammoniacal copper (I) chloride. A fresh sample of exhaust gas is then passed, as before, through this new solution. Here, the reduction in volume is the percentage volume of carbon monoxide present, because this solution absorbs carbon monoxide only.

(iii) *Measuring the amount of oxygen:* Finally, replace the boiling tubes of ammoniacal copper (I) chloride with boiling tubes containing alkaline pyrogallol solution. A fresh 100 cm³ sample of exhaust gas is passed through it. This time oxygen is the only gas absorbed. The decrease in volume therefore gives the percentage by volume of oxygen present in the exhaust gas.

(iv) *Measuring the amount of nitrogen:* The percentage by volume of nitrogen is determined by difference; i.e. by taking the total percentage of carbon dioxide, carbon monoxide and oxygen away from 100.

The following table shows the kind of results possible with this method:

Table 9. The analysis of car exhaust gases

Car — Ford Escort 1100, running cold, 4 star petrol

Gas	% by volume
Carbon dioxide	11.0
Carbon monoxide	3.5
Oxygen	8.4
Nitrogen (by difference)	77.1

On facing page A train leaving a coal-fired power station after making its delivery.

(12) Demonstrating the presence of tar, carbon dioxide, carbon monoxide and hydrogen sulphide in smoke from coal ***

What you will need

What you will need

An empty syrup tin with a push-on lid
A bell jar or similar container
A water pump (suction pump)
A 100 cm³ gas syringe
A bunsen or similar type burner
A pair of tin snips
A U-tube
Two rubber corks, one of which must fit the top of the bell jar
Two rubber bungs, with one hole, to fit the U-tube
Some glass tubing and rubber tubing
A tripod stand
Some litmus paper and lead acetate paper
Some glass wool
Some coal

What to do

What to do

Punch several small holes in the bottom of the syrup tin using a large nail. Using the tin snips cut a small flap in the side of the tin, about 3 cm square and about 3 cm up from the base. Make a hole in the lid large enough to fit one of the bungs. The tin now looks like a fireplace, the holes in the base being the "fire bars". Now assemble the rest of the apparatus as shown in diagram 14.

Place some coal on the "fire bars" and set light to it by holding the flame of the bunsen burner underneath. When the coal is burning switch on the suction pump. The smoke and tarry gases from the burning coal pass into the U-tube,

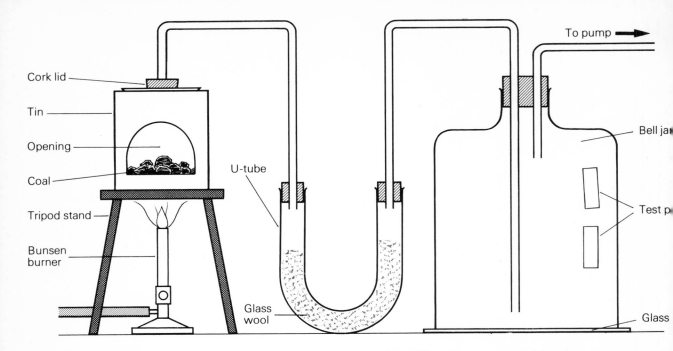

Labels (left to right):
Cork lid
Tin
Opening
Coal
Tripod stand
Bunsen burner
U-tube
Glass wool
To pump
Bell jar
Test paper
Glass

Diagram 14 Apparatus for detecting the by-products of coal smoke.

where the tar settles out. The smoke is drawn into the bell jar where its colour and composition can be seen with the naked eye. At the same time gases which cannot be seen, like carbon dioxide, also collect in the jar. Samples of the gases can be drawn out and analysed for carbon dioxide, carbon monoxide and oxygen as explained in experiment 11.

The results will vary as they do in real life because the amount of air supplied for burning, the type of coal, etc, varies.

Other tests

(i) The bell jar also shows that water vapour is present because water droplets can be seen forming on the cool sides of the jar.

(ii) If litmus paper and lead acetate paper are moistened and stuck to the side of the bell jar, they will change colour. This shows that hydrogen sulphide is present. It also indicates that there may be acid or ammonia fumes in the smoke.

(iii) Other fuels can also be tried. Sawdust and coke can be burned using the same apparatus. The smoke obtained can then be compared with the smoke from coal. Oil may be burned in a small glass dish placed inside the tin and resting on the "fire bars". Gas may be burned by placing the bunsen burner flame inside the tin and turning on the water pump.

In our experiments with these fuels the following results were obtained:

Table 10. By-products of burning fuels

Fuel	Tar	Acidity	Hydrogen sulphide	Water vapour
Coal	+ + +	basic	+ +	+ + +
Coke	none	none	none	+
Oil	none	none	none	+ + +
Sawdust	+ +	acid	none	+ + +
Gas (natural)	none	none	none	+ +

+ : small amount or trace, + + : moderate amount, + + + : large amount

(13) The production of nitrogen dioxide by sparking plugs **

We have seen already that sparking plugs make nitrogen dioxide from air. This is only formed in small amounts, so although it is poisonous it is not a great danger. However, in sunny climates it causes hydrocarbon fumes from exhaust pipes to change and become highly lethal. This experiment shows how nitrogen dioxide is made.

What you will need
One car sparking plug
Two long lengths of insulated copper wire
A small bottle, 250 cm³ capacity with a bakelite top
An induction coil (free standing or the car induction coil)
Some damp litmus paper

What to do
Make a small circular hole in the screw on the bakelite top of the jar, so that the plug can be screwed into it. Screw the bakelite top and plug back on to the jar. Attach the plug to the induction coil using the wires, and switch on the current (diagram 15). A spark will jump across the gap in the sparking plug. After five minutes of sparking the air in the jar will have become a pale brown haze. The brown fumes present are nitrogen dioxide fumes. They may be recognized by the colour, by smelling *(very cautiously)*, or by the fact they will turn damp blue litmus paper red.

If you cannot get hold of an induction coil the same effect can be had by removing the high tension lead from the sparking plug in a motor car, and attaching the lead to the plug in the jar. An earth wire from the body of the car to the side of the plug is also required.

In one such test a faint brown tinge was visible in the bottle after the engine had been running for 75 seconds. The nitrogen dioxide was recognized by the pungent smell in the bottle; and the blue litmus turned red.

Diagram 15 Apparatus for making nitrogen dioxide using a sparking plug.

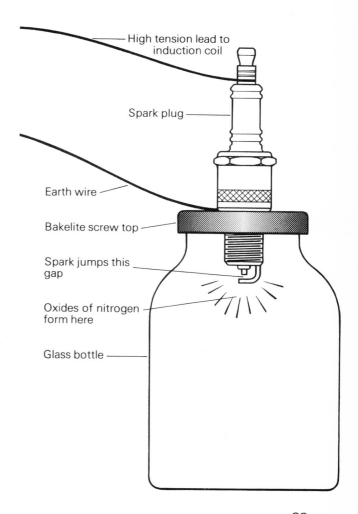

High tension lead to induction coil

Spark plug

Earth wire

Bakelite screw top

Spark jumps this gap

Oxides of nitrogen form here

Glass bottle

(14) Making artificial smog *

Smog is a mixture of smoke, invisible noxious gases and minute water droplets, which have been trapped near the ground. In normal weather the upper air is cooler than air near the ground. Any hot smoke and fumes released at near ground level will escape quickly into the upper atmosphere, since hot gases rise. But sometimes inversion takes place, and cold air is trapped near to the ground by a layer of warm air above it. This happens especially in winter, and on low ground and in river basins — where many large cities are built. In these conditions the smoke and noxious gases together with water vapour are trapped at ground level. Here they form a cloud full of dirt particles and noxious gases. Smogs are a serious health hazard and cause many deaths.

What you will need
A fish bowl or large jar
A sheet of thin aluminium foil
A little salt
Several ice cubes
Some paper tapers and matches

What to do
Set up the apparatus as shown in diagram 16. Light the taper, drop it into the bowl and close the bowl with the sheet of aluminium foil. Add the ice to the top of the foil and then sprinkle salt onto the ice. Watch what happens to the air in the jar.

Now remove the foil and ice/salt mixture. Wash out the inside of the bowl leaving a few drops of water clinging to the sides. The water makes the air in the bowl more moist. Steam from a kettle may be used for this purpose. Now light the paper taper again, and repeat the experiment. What happens to the air and smoke in the bowl now?

As the air in the jar cools the smog begins to thicken, becoming more discoloured. Eventually you will not be able to see through the jar. The demonstration will be just as successful if the paper is removed once the jar is full of smoke.

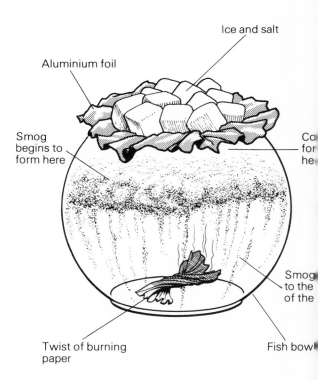

Aluminium foil

Ice and salt

Smog begins to form here

Co for he

Smog to the of the

Twist of burning paper

Fish bow

Diagram 16 Apparatus for making artificial smog.

(15) The cigarette machine **

When Sir Walter Raleigh first brought the smoking of tobacco into sixteenth-century England, he began a social habit which has lasted to this day. Modern medical research has shown that cigarette smoking is the main cause of several chest diseases including bronchitis. Other diseases such as lung cancer and heart trouble are also partly due to cigarette smoking over a long period of time. Smoking is a form of personal pollution best avoided.

What you will need
A U-tube with a side arm
Some cotton wool
A 0-100°C thermometer
Some rubber bungs, with one hole to fit the U-tube
Two conical flasks of 250 cm³ capacity
Some rubber bungs, with two holes to fit the conical flasks
Some glass and rubber tubing
A water pump
Some cockroaches or other insects
Some lime water

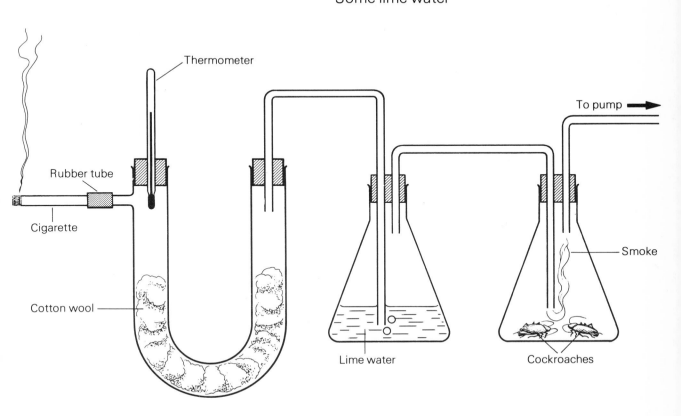

Diagram 17 The cigarette machine.

What to do

Set up the apparatus shown in diagram 17. Fill the U-tube with cotton wool and fix the water pump so that air can be drawn through the U-tube and the two conical flasks. Place lime water in the first conical flask and some cockroaches in the second (this may also be left empty).

Now light a cigarette and put it into the open end of the side arm of the U-tube. Switch on the water pump and watch how the smoke travels through the U-tube leaving behind a brown deposit. Can you identify this deposit? You will notice that the lime water has changed its appearance. Why is this? What has happened to the cockroaches?

Take some of the cotton wool which has a brown deposit on it and soak it in distilled water for several minutes. Now take a potted plant with some greenfly or aphids on the stems or leaves from a greenhouse or window sill. Spray or dip the leaves of the plants and the aphids in the solution of brown deposit from the cotton wool and distilled water. Watch what happens to the insects. What does this tell you about cigarette smoke?

During the experiment you should record the temperature of the incoming smoke on the thermometer. Does this temperature rise or fall as the cigarette gets shorter?

Table 12 gives the nicotine and tar contents of several well-known brands of cigarettes. In one experiment a "strong" variety contained enough nicotine to kill cockroaches, insects which are very hard to kill with common insecticides such as pyrethrum.

What the results show

(i) The poisonous chemicals present in cigarette smoke i.e. tar, nicotine and carbon monoxide.

(ii) The ability of nicotine to kill insects.

Table 12. Tar and nicotine content of cigarettes

Tar yield mg/cigarette	Brand	Filter or plain	Nicotine yield mg/cigarette
Under 4	Embassy Ultra Mild	F	under 0.3
7	Player's Mild De Luxe	F	0.3
11	Rothman's Masters	F	0.6
13	Player's Special Mild	F	0.8
15	Piccadilly No 7	F	0.8
18	Embassy Gold	F	1.2
19	Cameron	F	1.2
19	Rothman's King Size	F	1.4
20	Guards	F	1.3
20	Player's No 6 Filter	F	1.2
21	Kensitas Tipped	F	1.4
21	Sotheby's	F	1.4
23	Player's Filter Virginia	F	1.6
27	Woodbine Plain	P	1.7
31	Churchman's No 1	P	1.9
32	Gold Flake	P	2.0
38	Capstan Full Strength	P	3.2

(16) The effect of car fumes on germinating seeds **

What you will need
Three large bell jars with bung holes in the tops
Two rubber bungs, with two holes, to fit the bell jars
A water vacuum pump
Twelve small crystallizing dishes
A large plastic bag

Three sheets of glass on which to stand the bell jars
Some filter papers to fit into the crystallizing dishes
Some grass seed
A little Vaseline
Some glass and rubber tubing, and some Hoffman screw clamps

Diagram 18 Apparatus for testing the effect of car fumes on germinating seeds.

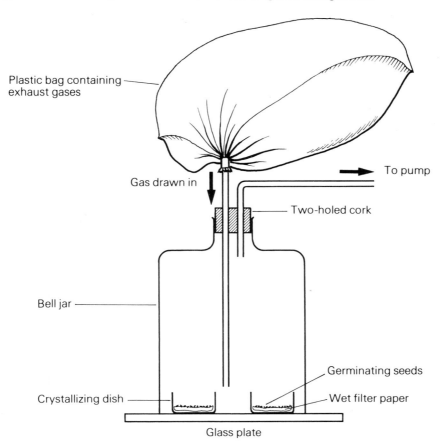

Plastic bag containing exhaust gases

To pump

Gas drawn in

Two-holed cork

Bell jar

Germinating seeds

Crystallizing dish

Wet filter paper

Glass plate

What to do

Place a filter paper in the base of each crystallizing dish and moisten it. Count out 50 grass seeds, and place on the paper in the dish. Smear the base of each bell jar with Vaseline and place one jar onto a sheet of glass to cover a set of four crystallizing dishes containing the seeds. Fit the two-holed rubber bungs into the bung hole at the top of the jar. Push in the glass tubing so that one length opens near the top of the jar and the other opens near the base among the crystallizing dishes (see diagram 18).

Place the mouth of the large plastic bag over the end of the exhaust pipe of a car (see experiment 11). Run the engine until the bag is full of exhaust gas. Seal the bag and return to the laboratory.

Using the water pump, draw clean air into one bell jar containing four dishes. Label this jar "control". Attach the plastic bag to the longer tube of the second bell jar and draw petrol fumes through the bell jar. In our experiment it took about 30 seconds to make sure that all the clean air had been removed from the jar. Seal off both tubes with the clamps. Label this jar "continuous exhaust fumes".

Now attach the plastic bag to the third bell jar and draw exhaust fumes into it. Label it "intermittent exhaust fumes". After twenty-four hours replace the exhaust fumes with clean air. Make sure that clean air is pulled through for at least one minute to remove all traces of exhaust gas. At the end of the second twenty-four hour period replace the clean air with exhaust gas once more. Continue

The motor car is certainly a boon to modern civilization but its by-products do untold damage to our plants, our wildlife and ourselves.

A Hungarian traffic policeman makes a spot check on the fume emission from a bus.

this change of atmosphere in the third jar throughout the experiment.

After one week remove the bell jars from the plates. Count the number of seeds that have germinated in each of the four crystallizing dishes under each of the three jars.

In one such experiment these results were obtained using grass seeds (perennial ryegrass S24):

What the results show

(i) Seedlings growing near roadsides may be badly affected by exhaust fumes.

(ii) Continuous exposure to exhaust fumes greatly reduces the germination rate for grass seeds.

(iii) Alternate exposure to air and exhaust gas hardly affects the germination rate.

Table 13. Percentage germination of grass seeds in exhaust fumes

Experiment	Time (days)	Control air	Intermittent air/exhaust	Continuous exhaust gas
1	7	80.0%	78.4%	27.2%
2	7	74.0%	76.0%	31.2%
3	7	91.0%	87.5%	45.5%
Averages		81.6%	80.6%	34.6%

III Natural forms of pollution

Introduction

The atmosphere is always being contaminated by natural pollution. Volcanoes have erupted from earliest times. They throw out millions of tonnes of dust and noxious gases into the air. Tornadoes and hurricanes tear at the earth's surface, throwing tonnes of dust into the upper atmosphere. Of course the dust finally settles, but in doing so it often destroys vegetation and animal life. We should look at such happenings as part of nature's processes.

In the same way, when plants and animals decompose after death, many noxious gases are given off. They are the by-products of the decomposition of carbohydrates and proteins by anaerobic bacteria. The gases given off are methane (natural gas), ammonia, hydrogen sulphide and others. Given time these by-products are removed by nature itself.

The difference between pollution by nature and pollution by humans lies in the strangeness of some of the chemicals which we produce.

Because of their strangeness, there are no natural processes able to remove them. What is more, the actual amounts of gases like carbon dioxide which we release into the atmosphere are often too large for nature to cope with. Modern industry is like a continuous volcanic eruption.

The following experiments show how nature produces various gases, some beneficial and some noxious.

On next page Diagram 19 Apparatus for producing carbon dioxide from rotting vegetation.

Below Volcanoes have erupted throughout history to pollute the atmosphere with noxious gases: Mount Etna erupting in 1852.

(17) Carbon dioxide from rotting vegetation **

What you will need
A water pump or aquarium aerator
Eight 100 cm³ conical flasks
Some fresh lime water
A strong caustic potash solution
A rotten apple or other decaying organic matter
Eight 2-hole rubber bungs to fit the flasks
Some glass and rubber tubing

What to do
Assemble the equipment as shown in diagram 19. You can see that the apparatus has two identical halves. One half acts as a control, the third flask having nothing on one side and grass cuttings or a rotten apple on the other side.

The air entering the flasks must first pass through the caustic potash solution. This removes any carbon dioxide. It then passes through lime water to make sure that all carbon dioxide has been removed. After passing over the organic matter the air enters the second lime water flask.

What the results show
Carbon dioxide is a colourless gas which turns lime water cloudy. It is heavier than air, and does not support combustion (this means that you cannot burn things in it). If carbon dioxide makes up more than 10% of the air in the atmosphere it is lethal to most animals, including human beings. Indeed the many deaths in fires due to suffocation happen because of the large amounts of carbon dioxide given off by burning materials.

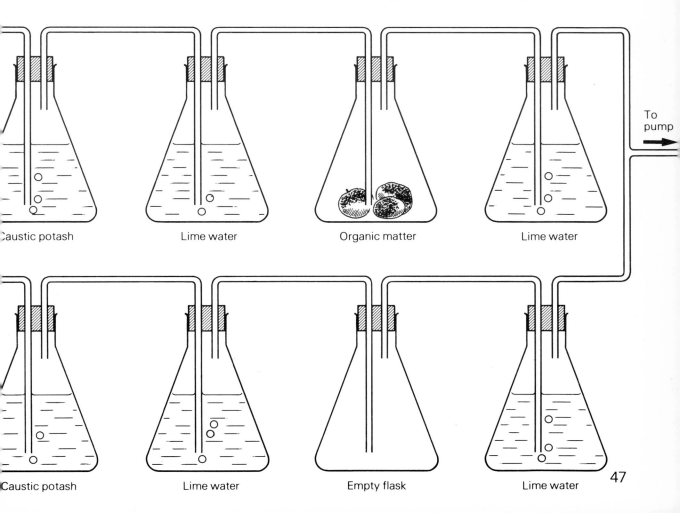

Caustic potash   Lime water   Organic matter   Lime water   To pump

Caustic potash   Lime water   Empty flask   Lime water

47

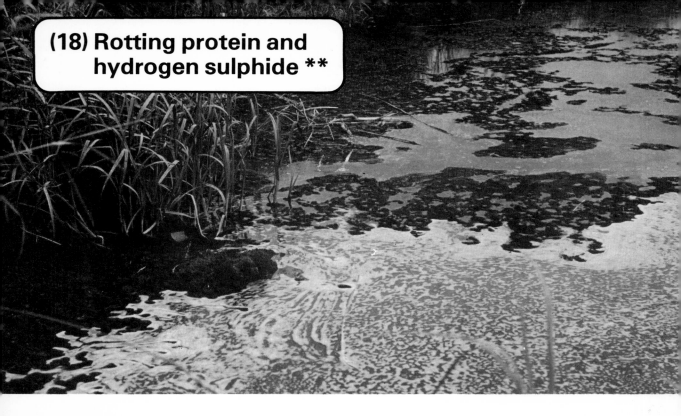

(18) Rotting protein and hydrogen sulphide **

Algae growing on a river in England, a sure sign that anaerobic bacteria are at work and that hydrogen sulphide is being produced.

Hydrogen sulphide is a gas with a smell which you can easily recognize — rotten eggs. This gas is given off by organic matter rich in sulphur compounds when they are decaying under anaerobic conditions.

Method 1
What you will need
Four eggs
A glass jar with a lid, e.g. a jam jar
Some lead acetate paper

What to do
Place the eggs inside the jar and leave them, undisturbed, for several weeks. Test the air inside the jar with a strip of damp lead acetate paper. This paper turns black in the presence of hydrogen sulphide.

Method 2
Another way to measure hydrogen sulphide pollution is to record the time taken for silver-plated objects (e.g. spoons) to turn black after cleaning. The silver reacts with tiny particles of sulphide in the air to form black silver sulphide.

What you will need
Two silver-plated spoons
Some silver cleaning solution and a cloth
A polythene bag

What to do
Select one cleaned spoon. Place it in a polythene bag and seal the neck of the bag. Leave the second spoon on a shelf and record the time taken for it to blacken. Compare it with the spoon in the bag. If possible record how long it takes for the spoon in the bag to blacken to the same extent. The difference in the two times is a good estimate of the extent of sulphide pollution in the air.

48

(19) The release of ammonia from pea seeds **

What you will need
A 500 cm³ flask or jar
Some cotton wool
About forty dried peas
Some red litmus paper

What to do
Put the peas in the flask. Add enough boiled and cooled tap-water just to cover them. Allow the seeds to soak for twenty-four hours and then pour away most of the surplus water. Make a cotton wool plug and push into the neck of the flask (diagram 20).

Place the flask and its contents in a warm spot near to a radiator and leave for fourteen days. During this time watch what happens to the seeds, keeping a written record.

At the end of the fourteen days open the jar and gently sniff the air inside. You should be able to recognize the pungent smell of ammonia (which is rather like smelling-salts). Further proof of the presence of ammonia may be got by placing a strip of damp red litmus paper in the air over the peas. The red litmus will turn blue.

What the results show
Vegetation rots naturally to give large amounts of a useful gas called ammonia.

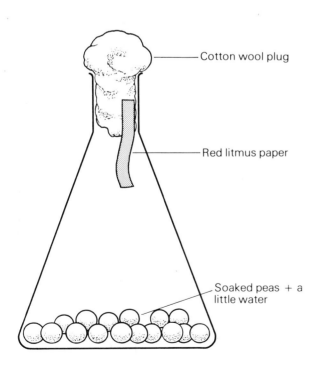

Cotton wool plug

Red litmus paper

Soaked peas + a little water

Diagram 20 Apparatus for producing ammonia from decaying pea seeds.

(20) Methane from sewage ***

All forms of animal dung, including human faeces, are decomposed by micro-organisms. When there is a limited supply of air a sort of fermentation type of breakdown occurs. When this happens methane gas is released instead of carbon dioxide. Methane gas is also known as marsh gas. Natural gas from the North Sea is also methane.

To produce methane from horse dung the following experiment may be used:

What you will need
A large plastic bottle or drum with a tight fitting lid (e.g. a sweet jar)
A one-hole rubber bung fitted with glass tubing
Some rubber tubing to fit the glass tubing
A little horse dung

What to do
Fill the drum or jar with the dung. Make sure it is well squashed down to remove as much air as possible. It is best to squeeze one layer before the next layer is added. At every stage make sure that the dung is kept moist. Insert the bung fitted with a glass tube and make sure that the glass tube is clear of the top of the horse dung. Place the drum in a warm place. The best temperature for the process is 36°C (blood temperature).

Diagram 21 Apparatus for producing methane from sewage

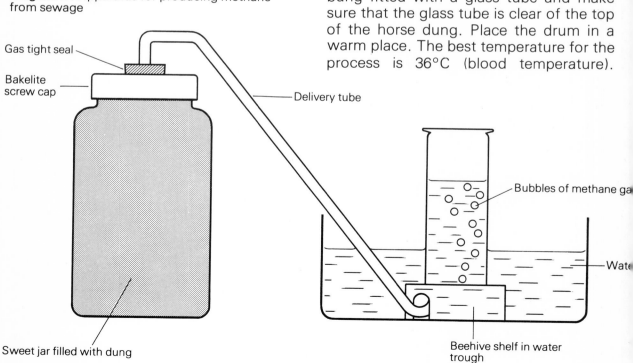

Gas tight seal

Bakelite screw cap

Delivery tube

Bubbles of methane ga

Wate

Sweet jar filled with dung

Beehive shelf in water trough

Leave for twelve days. At the end of this time methane starts to be given off. The gas is lighter than air and it may be collected over water (diagram 21).

Methane is a colourless gas; it has no smell and does not react with lime water. It burns with a blue, hot, clean flame.

In our experiment fourteen days were needed to make methane inside a glass sweet jar. With a plastic drum, methane production begins sooner because it is possible to squeeze out more of the air from the sewage in the first stages of packing.

A sewage works where the methane produced from the sewage is used to generate the electricity to power the works.

IV How the senses react to pollution

(21) The sense of smell **

The human sense of smell is weak compared to that of many other animals, such as dogs. Perhaps this is an advantage, considering the many pollutants released by industry — naphthalene, bromine, petrol fumes and other organic chemicals.

What you will need
A 50 cm³ wide-mouthed bottle
10 cm³ of oil of cloves

What to do
Put the oil of cloves into the jar and keep sniffing it. Record how long it takes before the smell disappears.

Leave the classroom for five minutes breathing deeply in the fresh air. Return to the room and sniff the oil of cloves again. Does it smell again now?

If it does, does this suggest that the sense of smell is weak but soon built up again by clean air?

What the results show
  (i) The sense of smell is sensitive at first but quickly loses its sensitivity.
 (ii) A short time in fresh air allows the sense of smell to recover.

(22) The sense of hearing **

In recent years there has been a big rise in the pollution of the atmosphere by noise. Sonic booms, pneumatic drills and jet aircraft, all add noise to the environment. Many youngsters attend discos where the noise level is dangerously high. The inner ear is a very delicate structure, and therefore it is not altogether surprising that ear damage and ear diseases are on the increase.

Unfortunately we lack a simple method for measuring how much we can hear. It is also hard to measure the noise made by things like trucks, motor-cycles and aircraft.

The following methods may be used to investigate hearing in people:

Method 1

What you will need
An ear muff or a large pad of cotton wool
A wrist watch with a loud tick

What to do
Cover one ear with the muff. With one ear open slowly walk away from the ticking watch. Test at what distance the tick of the watch fades away. It is important to test not only the distance of the watch from the ear, but also the direction it is in. In this way the range *and* direction of hearing may be tested.

Is one ear weaker than the other?

Is there a deaf area from which sounds cannot be heard?

How does human hearing ability compare with that of other animals?

Method 2

What you will need
One audio frequency generator
(The audio frequency generator should have a range from 10 cycles per second up to about 24 kilocycles per second)
One amplifier and loudspeaker

What to do
The person to be tested should sit facing the loudspeaker. Let the listener note the sound frequencies at which he first hears sound. Raise the frequency until one is reached which the listener cannot

This man is measuring the noise generated by *Concorde.*

hear. Now come down the range of frequencies, noting the one at which the listener hears sound again. Come down the scale until the listener shows that he cannot hear the note at the lower end of the scale.

Repeat the experiment for each ear in turn, making sure that the other is suitably masked.

Can you explain any difference in hearing between the two ears of the same person?

If several people of different ages are available, you may be able to correlate their hearing ability and age.

NOTE: In all these tests the loudness of the note given out should remain constant.

Another useful test is to try to correlate hearing ability with place where the person lives or works. For example, do people living on busy main roads have worse hearing than people living in quieter areas? Are car factory workers more "hard of hearing" than housewives?

Method 3

What you will need
A tape recorder

What to do
Record on tape a series of different noises. Record them in a random order.

The greater the variety of noises and sounds the better.

Play the tape to a group and get them to record their degree of pleasure on first hearing the sound. Use a three-point scale, pleasant, neutral, unpleasant.

Repeat the sounds but at a higher volume, and then at a lower volume. Is there any change in the degree of pleasure or displeasure on hearing the same sounds at different volume levels?

Change the frequency of the sounds by speeding up the tape recorder. This also changes the quality. What is the class reaction now?

Try to find out if there is any connection between volume, frequency and hearing response.

What the results show

(i) The effect of distance and direction on the ability to hear.
(ii) The range of sounds we can hear and tolerate.
(iii) The effect of getting used to continuous loud sounds.
(iv) Which sounds are pleasant, which are unpleasant and why.

Part of a poster issued by the Noise Abatement Society.

"NOISE IN ANY MACHINE IS A SIGN OF INEFFICIENCY

EXCEPTING IN THE CASE WHEN A NOISE IS THE AIM

SUCH AS A SHIP'S SIREN"

SIR JOHN THORNEYCROFT

(23) The sense of taste **

The first part of this experiment deals with finding which regions of the tongue are sensitive to saltiness, sweetness, sourness and bitterness.

The second part tests whether the breathing of smoke affects the sense of taste.

Part 1
What you will need
Four 50 cm³ beakers
Four glass rods each 7 cm long
Four labels
Five per cent sodium chloride solution (common salt)
Five per cent sucrose solution (common sugar)
Some pure lemon juice
Some cold strong tea (without sugar or milk)

What to do
Label the beakers *Salt solution*, *Sugar solution*, *Lemon juice* and *Tea* respectively. Collect in each beaker a small quantity of each liquid. In your laboratory notebook draw an outline plan of your tongue. Work in pairs, one person acting as subject, the other applying the following tests to the subject:
(a) With a clean glass rod place drops of the salt solution one by one over all parts of the surface of the tongue. The subject should inform his or her partner each time that he or she can taste the salt solution. The person applying the test should then plot, using the appropriate symbol (i.e. Sa for salt, Su for sugar, etc), these positions on the diagram of the subject's tongue.
(b) Rinse the mouth and repeat the test using the sugar solution.
(c) Again rinse the mouth and repeat the test with pure lemon juice.
(d) Rinse the mouth and repeat the tests with cold tea.

Where on the tongue are the taste points for salty, sweet, sour and bitter tastes mainly sited?

Part 2
What you will need
As for part 1 but with the addition of a cigarette (or a bell jar containing smoke from paper or coal or tobacco).

What to do
Place a drop of salt solution on the part of the tongue which can taste salt. Rinse out the mouth.

Now breathe in some smoke from a cigarette (or from the bell jar) holding the smoke in the mouth for some seconds and then exhaling. Add a drop of salt water to the same part of the tongue. In what way is the taste different if at all?

Repeat the experiment with sugar solution, pure lemon juice and cold strong tea. In what way do they taste different after contact with other substances?

Can you explain what is happening? What connection has this with atmospheric pollution?

What the results can show
(i) The location of the major taste areas on the tongue.
(ii) The effect of smoking and pollution of air on our sense of taste.

On next page Air polluting smoke from the Central Electricity Generating Board power station at Rickborough in Kent.

Conversion chart

To convert	into	multiply by
Atmospheres	Kilogramme/sq m	10,330
Atmospheres	Pounds/sq inch	14.7
Centigrade	Fahrenheit	(C x 1.8) + 32°
Centimetres	Feet	0.03281
Centimetres	Inches	0.3937
Centimetres	Metres	0.01
Cubic centimetres	Cubic feet	35.31 ÷ 1,000,000
Cubic centimetres	Cubic inches	0.06102
Cubic metres	Cubic centimetres	1,000,000
Cubic metres	Cubic feet	35.31
Cubic metres	Cubic inches	61,020
Cubic metres	Cubic yards	1.308
Gramme	Ounces (weight)	0.03527
Gramme	Pounds (weight)	0.002205
Gramme/sq m	Pounds/square foot	2.0481
Kilogramme	Gramme	1,000
Kilogramme	Pounds (weight)	2.2046
Kilogramme/sq m	Gramme/sq cm	0.1
Kilogramme/sq m	Pounds/sq inch (weight)	0.001422
Kilogramme/sq m	Pounds/sq foot (weight)	0.2048
Litres	Cubic centimetres	1,000
Litres	Cubic feet	0.03532
Litres	Cubic inches	61.03
Litres	Cubic metres	0.001
Litres	Gallons (Imperial)	0.2199
Litres	Pints	1.759
Metres	Centimetres	100
Metres	Inches	39.37
Metres	Feet	3.281
Metres	Yards	1.0936
Millimetres	Inches	0.03937
Square centimetres	Square inches	0.1550
Square metres	Square feet	10.764
Square metres	Square inches	1,550
Square metres	Square yards	1.196
Tons (2240)	Tonnes (1000 kg)	1.016

Glossary

Absorption The soaking up, or sucking in, of a liquid by a solid.

Acid A substance with a sour taste and which is often corrosive. It can be identified by the fact that it turns blue litmus paper red.

Aerate To put air, or any other gas into something. The apparatus used for this purpose is called an aerator.

Anaerobic Oxygen-hating.

Atmosphere The layer of gas which surrounds the earth.

Audio frequency generator An electrical device for making sounds of a certain pitch or **frequency.**

Bacteria Very small, microscopic-sized **organisms.**

Biology The science of living things.

Burette A glass tube marked with measurements of volume. It has a tap and is used to measure the amount of solution running out of it.

Carbohydrates Chemicals which contain carbon, hydrogen, and oxygen. They are an important source of energy for all living things.

Combustion Burning.

Concentration The amount of a substance in a given area or volume.

Contamination To spoil by touching or mixing with an unpleasant substance (see **pollution**).

Convoluted bone The rolled-up bone found in the ear.

Correlation A relationship between two factors where the action of one factor may bring about a reaction in its correlative.

Debris Bits of wreckage or rubbish.

Decomposition The breaking down of a chemical into simpler chemicals.

Dissolve To go into a **solution.**

Distilled water Water which has no **dissolved** solids in it. It is made by boiling and condensing ordinary water.

Pollutants in the air have eroded the stonework of Canterbury Cathedral.

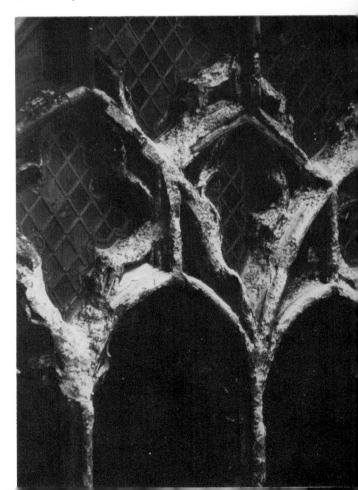

Element A substance which has not been combined with any other substance.

Environment This word has a lot of meanings nowadays; generally, it can be said to mean surroundings, or the world around us, although it is also applied to an animal's or plant's habitat or place where it lives.

Filter paper A material rather like blotting paper with tiny holes in it which allow liquids to pass through but trap solid materials.

Filtrate The clear liquid left after filtering.

Flow meter An instrument that measures the velocity of water or air flow.

Forceps An instrument, rather like a pair of tongs, used for handling small objects.

Fossil fuels Fuels which usually lie underground and which originate from dead plant and animal matter.

Frequency The number of times a particular event occurs within a space of time.

Germination When a seed first begins to grow shoots and roots.

Inversion An increase in temperature with height above the earth's surface. This is the reverse of the normal situation. It often occurs over large cities during special weather conditions in winter.

Lichen A fungus which lives on trees and rocks.

Lux Strength of light as measured by a light meter.

Membrane A thin piece of tissue which covers, lines, or separates, two different parts of the body.

Methane See **natural gas.**

Micro-organism A tiny single-celled **organism** which can only be seen with the aid of a **microscope.**

Microscope An instrument, rather like a telescope, which is used to produce an image which is larger than the actual object being viewed.

Mineral An inorganic (not **organic**) substance found in the ground.

Muffle furnace An electric oven, lined with fire bricks. It can be heated to a very high temperature of up to 1,000°C.

Natural gas A gas given off by decaying organic matter as it is eaten by **anaerobic bacteria.** It is also called **methane** or marsh gas.

Nicotine A chemical found in tobacco, and other plants. It is a natural insecticide.

Noxious Harmful, or unpleasant.

Organic Any chemical which is, or has been, part of a living plant or animal.

Organism Any plant or animal.

Parts per million The number of parts of a substance mixed with one million parts of another substance.

pH A measure of the **acid** content of a **solution**. The measure is on a scale of 0-14 where the neutral point is taken to be water at pH 7. A pH of less than 7 means high acidity while a pH of more than 7 means low acidity.

Photosynthesis The process by which green plants manufacture their food (**carbohydrates**) from the carbon dioxide and water in the atmosphere. Sunlight is essential for this process.

Plankton Microscopic animals found in water.

Pollution Contamination of an **environment** and disturbance of the natural balance in a harmful way.

Saturation The point at which a liquid is unable to **dissolve** any more of a substance.

Sewage A mixture of water and **organic** solids from houses and other places collected in large underground pipes called sewers.

Smog A harmful mixture of fog, smoke and other polluting gases in the **atmosphere.**

Solution A mixture of different substances. The term is usually used to mean the **dissolving** of solids in liquids e.g. the mixture of salt and water in sea-water.

Spores The tiny "seeds" of primitive plants.

Trachea The windpipe, or the air passage from the throat to the lungs.

Water vapour Steam, or water in the form of a gas.

The effects of air pollution from the Port Talbot steel works in Wales.

Further reading

D I Williams and D Anglesea, *Experiments on Land Pollution* (Wayland Publishers 1978)
D I Williams and D Anglesea, *Experiments on Water Pollution* (Wayland Publishers 1978)
D I Williams and D Anglesea, *Projects in Conservation* (Wayland Publishers 1978)

W A Andrews (ed), *Contours: studies of the Environment series* (Prentice-Hall 1972):
    A Guide to the Study of Environmental Pollution
    A Guide to the Study of Soil Ecology
    A Guide to the Study of Terrestial Ecology

R E Baker and J A Bushell, *The Unclean Planet* (Ginn and Co 1971)
R G Borden, *Sound Pollution* (University of Queensland Press 1976)
J Burton, *Pollution* (Blackie 1974)
Richard Carlyon, *What on Earth are we doing?* (Ladybird Books 1976)
Mike Lyth, *The War on Pollution* (Wayland Publishers 1977)
Richard Mabey, *The Pollution Handbook* (Penguin 1974)
A MacKillop, *Talking About the Environment* (Wayland Publishers 1973)
H F Norman, N Norman, A Triffett and D E Weiss, *Nature in the Balance* (Heinemann 1972)
Barbara Ward and Rene Dubos, *Only One Earth* (Deutsch 1972)

Index

Picture acknowledgements
The authors and publishers wish to thank the following for the pictures which appear in this book:

Heather Angel 15, 31, 48, 61; The British Steel Corporation *Frontispiece*; The Central Electricity Generating Board 56-7; Dawe Instruments Ltd 52-3; *The Manchester Evening News* 27, 44; P Morris (ARDEA Photographics) 59; The Noise Abatement Society 54; The Wayland Picture Library 11, 22, 37, 46, 51; D I Williams and D Anglesea 20, 23 (top and bottom), 29 (all pictures); The World Health Organization 7, 32, 33, 45.
Special thanks to Stan Martin for the illustrations on pages 9, 12, 13, 14, 16, 17, 18, 21, 24, 26, 30, 35, 38, 39, 40, 41, 43, 47, 49, 50.